Flying is ...

50 spotlights on the aviator life

by
Briska De L'Aven

cheerful
and
profound

1

Flying is ...

... being able to enjoy the
freedom above the clouds
with all your heart,
but without ever losing
touch with reality ...

2

Flying is ...

... happily escaping the rain at home, in both the literal and figurative sense ...

3

Flying is ...

... no longer unpacking your suitcase at home, but simply repacking it and happily living out of your suitcase all year round ...

4

Flying is ...

... only being able to enjoy your
private life at home in small doses,
but usually looking forward to a
break when you're on the road ...

5

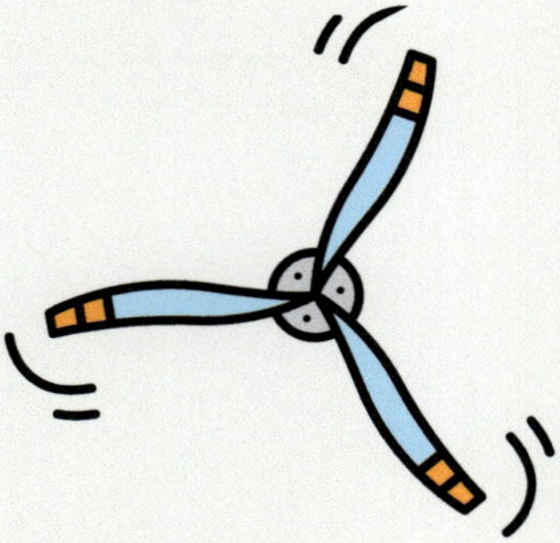

Flying is …

… getting up in the middle of
the night and at least
pretending to be as fresh
as a young morning …

6

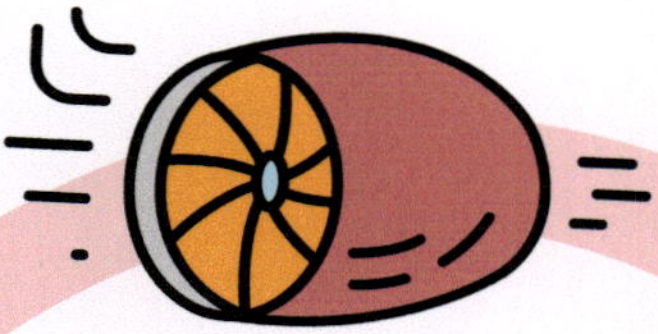

Flying is ...

... having to leave all your
heartache behind at home ...

7

Flying is ...

... having lots of opportunities
on the road, learning a lot
and using these opportunities
creatively and enthusiastically ...

8

Flying is ...

... intensively training serenity and
patience and simply no longer
being surprised by anything
in the air or on the ground ...

9

Flying is ...

... freezing in an ice-cold plane
at 5 a.m. in winter and still
not putting on an icy face ...

10

Flying is ...

... realizing at some point that
the years spent travelling have
flown by and hoping that you
haven't missed too much at home ...

11

Flying is ...

... sometimes feeling very ambivalent about the phone when you're on call and would rather stay at home ...

12

Flying is ...

... being far away from home
at Christmas, yet not feeling
lonely at all and making
the best of everything ...

13

Flying is ...

... being able to discover
interesting and beautiful things
while exploring the world alone
or with some crewmembers
and enjoying life ...

14

Flying is ...

... having the ability to work
harmoniously with complete
strangers without prior notice
and often for days on end
in a very confined space ...

15

Flying is ...

... to act with usual aplomb
even under extreme stress,
but above all appearing as
confident as possible ...

16

Flying is ...

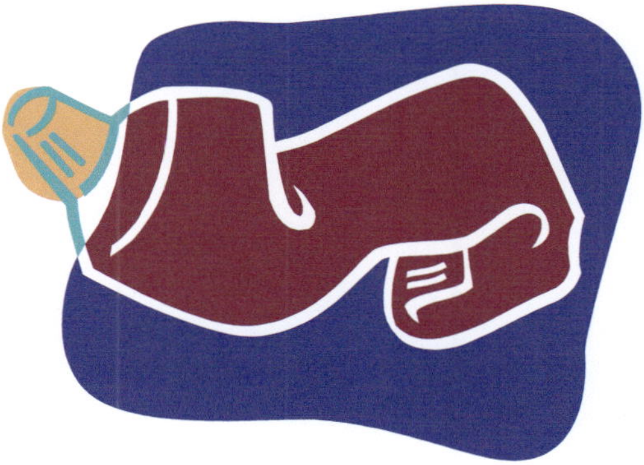

... sometimes feeling like a squeezed tube, but still managing to get something out of it ...

17

Flying is ...

... being able to become addicted
to it and, after weighing up
all the pros and cons,
coming to the conclusion
that flying is simply great ...

18

Flying is ...

... a technically quite perfectly organized thing, and it is more important than ever to put people at the centre of attention ...

10

Flying is ...

... when dealing with passengers
and crew, every now and then
hitting the bull's-eye
and being happy about it ...

Flying is ...

... anticipating every eventuality at all times and handling rare emergencies as if they were the most commonplace and normal thing in the world ...

Flying is ...

... quickly establishing a good rapport with many crewmembers and being able to talk amazingly openly about all kinds of things with each other ...

Flying is ...

... always having to plan far
ahead into the future,
but still finding the roulette of
days off and flight assignments
exciting every time ...

25

Flying is ...

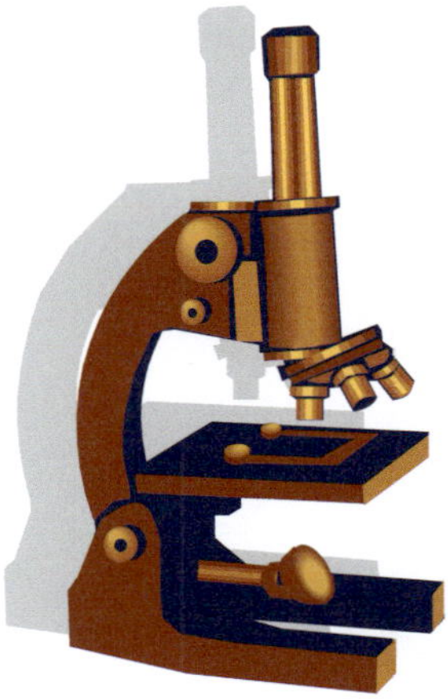

... always being on display and
bravely stepping into the spotlight,
even when you feel like
you'd much rather crawl into
a mouse hole ...

24

Flying is ...

... to know that our colourful world
has now merged into a single
megacity and that we cannot
all enjoy life on our own on
a lonely island of bliss ...

25

Flying is ...

... getting the curve right in time and, when the situation requires it, being able to execute a necessary 180-degree turn with lightning speed and precision ...

Flying is ...

... seeing clearly and being
able to make quick decisions ...

27

Flying is ...

... being in a festive mood during layover, but still managing to be totally fit for the next flight shift ...

Flying is ...

... always putting on a friendly face, behind which private feelings are hidden, and always exercising self-control and the necessary discipline ...

Flying is ...

... being able to work at top speed without losing track and, above all, never staying on the open road ...

Flying is ...

... taking care of your own physical and mental well-being while travelling, so that your muscles and nerves do not tire prematurely ...

31

Flying is ...

... a challenge for love and
a great opportunity to
practice mutual understanding,
letting go, and trust ...

32

Flying is ...

... having the right tools for all possible problems on board and being able to use them safely ...

Flying is …

… like a piece of cake:
it often tastes great,
but at times it can also be
very hard to digest and
sit heavy in your stomach …

Flying is ...

... moving through the world with curiosity, unobtrusively and without prejudice, sending positive signals in this way ...

Flying is ...

... admiring the treasures of this world with great interest, lovingly embracing the gigantic colourful diversity in our hearts and enjoying it with enthusiasm ...

Flying is ...

... being attentive to the concerns
and needs of fellow travellers,
so that the time spent together
in this flying tube is as pleasant
as possible for everyone ...

37

Flying is ...

... viewing life from above, frequently changing your perspective, and learning to appreciate and love the eternal dance between distance and closeness ...

Flying is ...

... perhaps no longer believing
in Santa Claus, but not giving
up faith in the good in people ...

Flying is ...

... giving your fellow travellers
as much space as possible,
which is important for their ego,
without giving up the space
you need yourself ...

40

Flying is ...

... finding the right mix of intense communication and solitude, and consciously strengthening your inner balance again and again ...

41

Flying is ...

... always trying professionally to
get all the balls over the net
and, depending on the situation,
keeping them as flat as possible ...

42

Flying is ...

... in everyday life from the
inside shining like a hollowed-out
and illuminated pumpkin
all year round ...

Flying is ...

... always being a beacon for
fellow travellers that they
can trust and that
safely lights their way
in all emergencies ...

44

Flying is ...

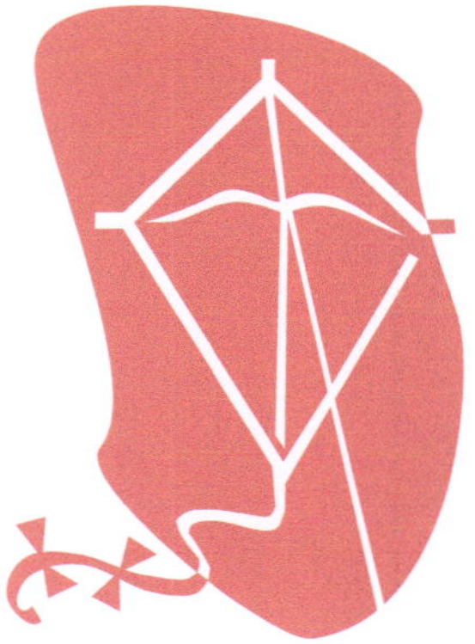

... also fun and games,
but you should master
all the rules of the game,
otherwise fun could
unexpectedly turn serious ...

45

Flying is ...

... knowing that others also get worked up sometimes and being very understanding about it ...

Flying is ...

... not avoiding certain people,
but graciously accepting
their presence ...

47

Flying is ...

... recognizing the qualitative
difference between an irregular
and an unregulated life
and acting accordingly ...

48

Flying is ...

... and remains an adventure,
even though aviation has
become highly automated,
commercialized, and
a mass phenomenon ...

40

Flying is ...

... to view one's own origins from
ever-changing perspectives and
recognize that every person
lives in their own vast universe
within the depths of space
that we all share ...

Flying is ...

... the best proof that only PEACE
can take us everywhere and
help us move forward, both
literally and figuratively ...

Briska De L'Aven

flew as a flight attendant on
B727/B737/B747-200/400/
AVRO RJ 85/CRJ-700/900/
Embraer E190/E195/
purserette B737
and was an AVRO/CRJ emergency trainer.

From April 1978 to the end of 2020,
with a break from flying between
the end of 1992 and 2003.

Briska also flew for several years as
a private pilot on a small single-engine
Cessna 150/152 in Germany and the
USA/Florida.

She looks back on over three decades
with many thousands of short, medium,
and long-haul flights.

Cheerful and profound,
Briska creatively illuminates the
many colourful sides of flying
from all kinds of angles.

IMPRINT

Bibliographic information from the German National Library: The German National Library lists this publication in the German National Bibliography; detailed bibliographic data is available online at www.dnb.de.

Brigitte Meyer-Simon Verlag
Am Blankenmoor 11a / D-38518 Gifhorn
www.leben-trifft-dich.com

Design Werner Liebchen,
www.liebchendesign.de

Printing: Libri Plureos GmbH
Friedensallee 273, 22763 Hamburg
ISBN: 978-3-930965-50-2

Picture credits:
p. 3: Vecteezy, Hemaraj Laten, p. 15: Vecteezy, Husein Signart, p. 17: Vecteezy, Deni Subianto, p. 19: Vecteezy, Lia Aramburu, p. 25: Vecteezy, abby-design, p. 29: Vecteezy, Matt Cole, p. 39: Vecteezy, Muhammad Adnan, p. 43: Vecteezy, Nix Art, p. 51: Vecteezy, Giuseppe Ramos, p. 59: Vecteezy, Johan Isevind, p. 67: Vecteezy, Pisut Tardging, p. 81: Vecteezy, Deni Subianto, p. 97: Vecteezy, Adipra Studio, p. 99: Vecteezy, Ilhustrator21, p. 101: Vecteezy, Diana Johjanna Velasques, Airplane Icons: Vecteezy, Zeybart and Vecteezy, Mehedi Hasan Salim